REDA EL-MERNISSI

# Chemistry didactics and ICTE: Towards a digital pedagogy

REDA EL-MERNISSI

# Chemistry didactics and ICTE: Towards a digital pedagogy

## Didactics of chemistry and ICTE: Towards an interactive digital pedagogy

ScienciaScripts

**Imprint**
Any brand names and product names mentioned in this book are subject to trademark, brand or patent protection and are trademarks or registered trademarks of their respective holders. The use of brand names, product names, common names, trade names, product descriptions etc. even without a particular marking in this work is in no way to be construed to mean that such names may be regarded as unrestricted in respect of trademark and brand protection legislation and could thus be used by anyone.

Cover image: www.ingimage.com

This book is a translation from the original published under ISBN 978-620-6-72304-2.

Publisher:
Sciencia Scripts
is a trademark of
Dodo Books Indian Ocean Ltd. and OmniScriptum S.R.L publishing group

120 High Road, East Finchley, London, N2 9ED, United Kingdom
Str. Armeneasca 28/1, office 1, Chisinau MD-2012, Republic of Moldova, Europe
Managing Directors: Ieva Konstantinova, Victoria Ursu
info@omniscriptum.com

Printed at: see last page
**ISBN: 978-620-8-40570-0**

SCIENCE
EDUCATION
Dr. REDA EL-MERNISSI

# Didactics of chemistry and ICTE: Towards an interactive digital pedagogy

## Dedication

*With the expression of my gratitude, I dedicate this modest work to those for whom, whatever the terms embraced, I could never manage to express my sincere love.*

*I dedicate this book to the soul of my dear father AYAD EL-MERNISSI. Words cannot express the pain I have felt since you left. Every day I think of you and feel incomplete without you. Your departure has left a void that no one can fill. May God have his soul in his mercy, and forgive him.*

*To the woman who suffered without letting me suffer, who never said no to my demands and who spared no effort to make me happy: my lovely mother CHAMA EL-HADDAD.*

*To my wife RADIA ES SEBBANI, who has never ceased to help, encourage and support me throughout my studies. May God protect her and grant her health and happiness.*

*To my flowers Roeya and Niema, I ask God to help me educate them and guide them on the right path.*

*To my dearest brothers and sisters, thank you for always being by my side with your presence and your love, giving taste and meaning to our family life.*

*To my friends and colleagues, For their encouragement and for all the good times we've had together. I hope our friendship will last forever.*

*To all those who love me and whom I love.*

*To all those who have helped me in any way*

## General introduction

Our research book is in the field of chemistry didactics. We aim to enhance the integration of ICTE into the teaching process in the case of the lesson entitled "Spontaneous transformations in batteries and energy harvesting: Indirect electron transfer". In recent times, we've noticed a special craze for the use and integration of ICTE in chemistry teaching. This strategy has become an integral part of the daily life of the teacher in a qualifying high school.

Modeling or simulations enable concepts or phenomena to be "visualized or shown" on a nanoscopic scale, generally making it easier for learners to understand. ICTs are like "magic" microscopes, allowing students to take journeys into the world of atoms and molecules, stimulating their creativity, imagination and learning. Learners, in turn, have demonstrated their motivation for activities taught with the help of information and communication technologies, and teaching using ICTE tools encourages learners and teachers to appropriate and master a dual practice: practical teaching activity and mastery of technological tools.

However, this book presents the results of work we carried out with 10 science classes (SVT, PC, and SM) in two Moroccan high schools. The aim of this work is to test the impact of integrating ICT into the teaching of science and chemistry at secondary level in Morocco, and also to answer a questionnaire filled in by students who benefited from an experience alone, and another group who benefited from the training of an experience and also to these tools from the ICT program.

# Author

**Dr. Reda El-Mernissi**

# List of abbreviations

| | |
|---|---|
| **SITE** | The International Society for Technology in Education |
| **ICT** | Information and communication technologies |
| **SVT** | Life and Earth Sciences |
| **PC** | Physics and Chemistry |
| **SM** | Mathematical Sciences |

# Table of contents

# Introduction

Numerous studies have described the barriers to technology integration in education. Research has grouped these barriers into two broad categories: first-order barriers, linked to external factors such as access to technology and equipment, and second-order barriers, linked to internal factors such as the beliefs and skills of school members. Over the past decade, as schools have become increasingly equipped with digital technologies, the focus of research has shifted from first-order to second-order barriers. The results of older studies examining the relevance of barriers to technology are difficult to apply to today's situation, and the exact interaction of these types of barrier remains unclear.

## 1. Integrating technology into science classes

Various studies have defined and measured technology integration, also referred to as technology adoption or assimilation in education [1]. Depending on the purpose of the study, technology integration can be conceptualized at a quantitative or qualitative level [2]. A focus on the quantitative aspects of technology integration provides an overview of the descriptive frequency of technology use, while the qualitative focus of technology integration includes normative measures such as efficiency, alignment with pedagogical practices or effectiveness of technology use for teaching and learning [2]. Quantitative measures of technology integration seem straightforward enough, but methodological problems can affect their application. For example, an estimate of the average number of weekly minutes a student spends using a computer or the Internet has been used as a rate of educational technology use [3]. However, global estimates of time can be difficult to estimate, and it is difficult to assess fluctuations in frequency with a global score. For this reason, some studies have incorporated more global and less precise estimates that focus on the relative percentage of time spent using ICT rather than exact minutes: in a recent large-scale European survey of teachers, for example, overall use of digital technologies was measured using separate ratings of the percentages of ICT use by students and teachers in lessons [4]. Items relating to the frequency of use of specific technical devices or the frequency of use of tools for certain purposes have been condensed for further analysis. However, these scales can be difficult to interpret, as information on the frequency of use of single tools is mixed with information on the variety of tools or the purpose of the tools. There are problems have been encountered with survey instruments that attempt to assess qualitative aspects of technology integration, such as making student learning more active or helping students solve problems [5], in addition, the aims of technology integration have been assessed as the use of technology for student-centered activities, and the facilitation of high-level thinking tasks [6]. Other studies have provided rather broad assessments and assigned rather broad and global scores to comprehensiveness and confidence in technology integration for educational innovation and attempted to correlate with student outcomes [7].

## 2. Obstacles to integrating ICTE into teaching and learning

Several theoretical models have been developed to identify and clarify the key factors that promote and hinder the use of technology in the classroom by teachers and students. Models that aim to explain technology use can be classified into two distinct categories: models that identify obstacles to technology use and focus on barrier removal, and models that highlight enabling factors and skill accumulation [8]. Studies have shown that removing barriers affects not only the use of these technologies by teachers, but also the use of computers as learning tools by students [9]. Research into the factors hindering computer use has distinguished between barriers, such as the lack of hardware and software called first-order barriers, and deeper barriers, such as the lack of teacher skills called second-order barriers [10]. This proposed categorization of barriers has been empirically tested and confirmed by several researchers[11], first-order barriers, such as the availability of digital devices, have been reduced in recent decades, but internal barriers, such as teacher skills still exist and are more difficult to overcome. Recent studies in the USA seem to confirm this hypothesis. However, the barriers addressed vary considerably. [12] identified attitudes and beliefs towards technology, as well as current levels of knowledge and skills, as the most significant barriers to teachers' use of technology. In qualitative interviews, [13] found that external barriers still exist, but overcoming internal barriers is more relevant, and another study showed that the comfort of IT functions (second-order barrier) was one of the strongest predictors of teachers' intention to use IT averages in schools [6]. Studies have shown that solutions to first-order barriers significantly affect, both directly and indirectly, second-order barriers, in order to activate teachers to use technology in higher and lower tasks [11]. In summary, first- and second-order barriers were found to impact the use or integration of technology in the classroom. Generally, the evidence indicated a trend towards increasing importance of second-order barriers, and first-order barriers could still be a major problem, as the exact interaction with these types of barriers is not yet clear.

## 3. How to improve teachers' skills

The need for teachers to use technology to improve both teaching practices and student learning has become increasingly apparent. One of the main ways to help teachers use technology more effectively is through professional development. Professional development is offered to practising teachers to improve their knowledge, strategies and other characteristics that can positively influence their teaching [14]. Empirical evidence has shown that many professional development programs are effective in improving the way teachers integrate technology into their classrooms; teachers who benefit from quality professional development are more likely to see their skills and abilities improve [15]. Given recent evidence of the importance of value beliefs in how teachers ultimately integrate technology, researchers have called for professional development programs to also address teachers' value beliefs regarding the use of instructional technology [16]. However, few studies to date have responded to the call to examine the relationship between professional development exposure and teachers' value beliefs about technology. A better understanding of this topic would support efforts to target value beliefs as an outcome of professional development. In addition, a more integrated examination of the relative influence of professional development exposure on teachers' perceived ability and value beliefs offered a more holistic picture of how professional development programs can help teachers use technology more consistently and purposefully. In order to design effective and targeted professional development programs, it is necessary to understand the possible outcomes on teachers, and it is necessary for institutions to invest in professional development programs, and further research is also needed to understand the different factors likely to influence the effectiveness of these efforts.

## 4. How professional development improves technology integration

The aim of technology professional development programs is to help teachers use technology more consistently and purposefully. However, researchers have noted that the ways in which exposure to professional development are often indirect, mediated by variables such as perceived ability and subject matter beliefs [17]. Effective technology integration begins with teacher change, therefore, professional development programs aimed at changing the way technologies are used in the classroom and also should target teacher-related factors [18], This involves not only improving teachers' strategic and pedagogical knowledge, but also their perceived abilities and attitudes. Empirical research has also supported the mediating role of perceived ability and value beliefs. For example, in an in-depth study of 1,382 practising teachers, Inan and Lowther found that an increase in teachers' technology skills was associated with gains in perceived ability and value beliefs [19], according to the results of their study, ability and value beliefs may also mediate the relationship between technology skills acquired through professional development and teachers' technology integration practices, with these variables having a direct influence on the extent to which teachers integrate technology into the classroom. Studies have specifically shown how exposure to professional development can contribute to and improve teachers' use of technology. The goal of technology professional development programs, and technology integration efforts in general, is to leverage available technology tools to enhance student learning. The International Society for Technology in Education (SITE) has established standards for educators that provide guidelines for encouraging learners.

The effectiveness of professional development programs is often measured by the amount and type of technology used by teachers [20], and researchers tend to talk about what tools teachers use and how often they use them. Indeed, the ultimate goal of teacher professional development in technology is to help teachers master the integration of technology to improve student outcomes (SITE, 2020), professional developments influence student learning [21], and the way technology is integrated facilitates the learning process [22]. It is clear that teachers develop quality learning experiences enriched by technology that enable students to develop their cognitive skills. A review of the literature revealed that fewer studies examined how students used technology to engage in cognitive learning tasks. Some studies used qualitative data, such as student interviews and observations [23], while others investigated students' learning engagement in digital environments, but used qualitative data [24]. It is therefore useful to consider a framework from which a measurement tool could be developed.

## 5. Thematic analysis

The integration of ICT into secondary education is characterized by common themes, patterns, distinctions and variations from subject to subject. The studies reviewed emphasize that ICT integration is not only an essential tool for making teaching more interesting and sufficiently imparting 21st century skills to learners in today's secondary schools, but also plays a vital role in improving learning outcomes, motivation, engagement and the development of students' critical and innovative thinking [25]. The studies reviewed emphasize that ICT integration is not only an essential tool for making teaching more interesting and sufficiently imparting 21st century skills to learners in today's secondary schools, but also plays a vital role in improving learning outcomes, motivation, engagement and the development of students' critical and innovative thinking [13], [26]. Indeed, several factors have influenced and impacted on the successful integration of ICT in various subjects, these factors can be student-related and teacher-related factors, student-related factors are those that affect students' learning processes, while teacher-related factors hinder the teacher's ability to provide learning experiences as part of the teaching process. Studies have shown that student-related factors can be minimized by integrating appropriate ICT tools, particularly tools and methods that align with learning objectives, subject content and pedagogy [27], [28]. For ICT integration to be effective, a number of considerations need to be taken into account, such as the quality of ICT tools and methodologies, student involvement, learning interaction and the assessment methods to be used [13] (Schmitz et al., 2022). On the other hand, teacher-related factors, such as inadequate professional development, lack of sufficient academic support and pedagogical knowledge, influence teachers' attitudes, beliefs and behaviors, which determine their desire to integrate ICT into different subjects and their effective integration.

## 6. ICTE applications in education

Long confined to the realm further afield than educational systems, ICT first appeared in university education before finding its way into high schools, the Internet, etc. ICTE has long been recognized as a means of demonstrating experiments that are costly, difficult to perform, dangerous, or to compensate for lack of equipment [29]. The use of technology in the classroom has both beneficial and potential implications. Educators and learners should see it as an opportunity to overcome obstacles to achieving their goals [30]. Most of the students interviewed in several surveys acknowledged that the use of ICTE in lessons was beneficial and helped them to think more critically and provide more reasoned answers in exams. Because ICTE enables students to complete assignments at different times, due to factors such as autonomous learning and individual differences [31], it is considered a powerful instrument that can help improve education in a variety of ways, including making it easier for teachers to create teaching materials and providing new methods of learning and education. . It will therefore be up to instructional designers and educators to use the potential of advanced digital technology to revolutionize education, so that effective and efficient education is available to everyone, everywhere [32], [33]. Technology has continued to play an essential role in teaching children, with digital learning fostering creativity and giving students a sense of achievement, encouraging further learning by moving away from traditional techniques. All nations have been able to adopt distance learning technologies using a combination of TV, radio, online and mobile platforms, which is commendable, these technologies allow easy access to information, easy retention, increased storage, and improved presentation of information, for this education has become more interactive, knowledge sharing easier and increased enthusiasm for learning[34], [35], [36]. With the advent of new technologies, teachers must learn to use a variety of computer technologies in teaching, such as smartphones and tablet computers, or risk being marginalized. Teachers must also exploit all the resources available online to ensure that their courses are lively, engaging and up-to-date. Technology is not limited to video games and animated film viewing; the benefits are determined by how students, parents and teachers use these techniques, as technological innovations have helped to facilitate learning across all age groups and subject areas. When technology is used effectively for pedagogical purposes, the overall educational experience improves, and to facilitate access and accelerate the adoption of digital learning, it is important to make e-learning systems compatible with new smart devices.

# Application

# Didactics of chemistry and ICTE: Towards an interactive digital pedagogy

Reda El-Mernissi[1] , Abdelali idrissi Taouarti[1]

[1]Direction provinciale de Meknès, l'académie régional de l'éducation Fès-Meknès, Minister of Education.

## Summary

Information and Communication Technologies (ICT) play an important role in education and other fields. ICT has become an important pedagogical tool in the training and development of students, both in terms of teamwork and individual skills. The growing use of simulation in teaching raises a number of persistent questions about the mechanisms involved in learning. The aim of this article is to study the impact of the pedagogical use of simulation software on the educational, learning and teaching process, and to demonstrate the effectiveness of simulated situations in a course entitled "Spontaneous transformations in batteries and energy production" for 384 2nd-year baccalauréat students in 10 science classes, from different options. The results obtained showed that the use of simulation software in the classroom enables teachers to improve the quality of chemical science teaching, make up for the lack of practical work, and enable teachers to solve complex situations.

***Keywords:*** ICT: Information and communication technologies; teacher; learner.

# Introduction

Since the adoption in 2005 of the strategy for integrating information and communication technologies (ICT) into Moroccan public education, teachers of chemistry at high school level have been strongly encouraged to integrate and use ICT as a pedagogical tool in their teaching practices. The general use of ICT in various areas of the public and private sector, even if not reaching the required level, is accelerating. The field of education is one of those benefiting from these IT tools [37], [38], [39], and strategies and efforts have been undertaken by the government to promote the use of ICT in education in general. Simulation applications in the school environment can activate multiple skills in chemistry students to formulate hypotheses and interpret results [40], [41], [42].

Simulation techniques are called in silico methods or computational chemistry methods [43], and are most widely used in scientific research [44], and more specifically in the design of new anti-cancer molecules [45], such as colorectal cancer [46], and breast cancer [47]. In silico methods encompass a very wide range of numerical methods based on the laws of physics and chemistry, such as molecular docking (El-Mernissi et al., 2022), the ADME pharmacokinetic filter [49], and dynamic simulation [50], which can be used in drug discovery in general [51].

Our aim in this article is to show the importance and potential of taking advantage of the benefits of computer simulators in high school teaching, in order to overcome and solve the problems associated with the complexity of carrying out practical work in the laboratory, and consequently simplify and concretize chemistry phenomena in high school students by integrating simulators into their teaching practices.

## Context

The use of ICT in teaching has many advantages, such as promoting interactive learning, presenting courses simply and effectively, facilitating interaction between teacher and learner in any complex situation, improving motivation and optimizing performance. Today, it has become necessary for the modern teacher to be trained in those skills that match the demands of ICT. In addition, the lack of scientific equipment in school science laboratories, and the absence of preparators play a negative role in the development of manipulative skills, the fostering of knowledge learning, student motivation, learning to work independently, and working in groups[52].

In Morocco, studies in 2015 showed that 68% of teachers only carry out less than 50% of the experiments programmed in the textbook [53], but now this number is bound to increase. To solve this problem, the Moroccan Ministry of National Education has launched a "Nafida 2" program, to promote access to information and communication technologies for the teaching family as part of the 2018-2028 ten-year plan.

Our interest in this digital activity in high schools is mainly motivated by proposals that information and communication technologies (ICT) can be used to enhance learning among learners.

In this article, we'll answer a number of questions:

- What is the impact of ICT use on chemistry practices and learning in general? What impact does the use of ICT have on learners' learning after the experiment? What is the impact of ICT use on student learning, analysis and problem solving?

We're going to answer these questions, based on the results of our research into the effects of ICT on the learning of high school chemistry. In doing so, we will demonstrate the integration of experimentation and ICT in high school teaching.

## I. Materials and methods

In this study we present the results from 10 classes in the 2022-2023 and 2023-2024 school years. These students were divided into two sections; a first section benefited only from the experiment "Spontaneous transformations in batteries and energy collection: Indirect electron transfer", these classes are: 1 science math classes (M1), 2 life and earth science classes (V1 and V2), and 2 physical science classes (P1and P2), then the questionnaire is filled after the experiment, the second section contains: 1 math science class (M2T), 2 physical science classes (P3T and P4T), and 2 life and earth science classes (V3T and V4T), these pupils benefited from the experiment and also the use of information and communication technologies (ICT) in the same lesson, then the same questionnaire was filled in. The schools involved in this study belong to two different establishments: the Lycée Ouad Elmakazen and the Lycée Anassi1, both belonging to the Fès-Meknès regional education and training academy. Students in the first section completed the questionnaire after the experiment, but students in the second section completed the questionnaire after the experiment and the integration of (ICT), the use of both methods are intended to test the performance of these techniques on learners' learning.

### 1. Handling

The lesson is entitled "Spontaneous transformations in batteries and energy collection: Indirect electron transfer". Figure 1 shows the four essential experimental steps for this lesson (A, B, C, and D).

Fig (A) is the starting point: 50ml of copper(II) sulfate solution, concentration C = 0.1mol/l (blue in color), and a Cu(**s**) copper plate are introduced into beaker A.

In a beaker B, 50ml of a zinc (II) sulfate solution of concentration C = 0.50 mol/l (colorless), and a zinc plate Zn(**s**) are introduced.

In figure (B): a freshly stripped copper Cu(s) plate is immersed in a copper(II) sulfate solution (beaker A), and a freshly stripped zinc Zn(s) plate is immersed in zinc(II) sulfate (beaker A). Allow to evolve for a few minutes, then observe the system.

In figure (C), the beakers of the two solutions (A) and (B) are connected using filter paper soaked in an aqueous solution of potassium chloride $K^+$ (aq) + $Cl^-$ (aq) (salt bridge).

In figure (D), electrical wires, an ammeter and a switch K are connected in series between the two plates. The COM pole is connected to the Zn(s) plate, and the A pole to the Cu(s) plate. Switch K is closed, and the ammeter shows that a current is flowing through the circuit at 0.76mA. Leave the experiment for a while, bring out the two plates and observe their states. Figure 1 below shows the steps involved.

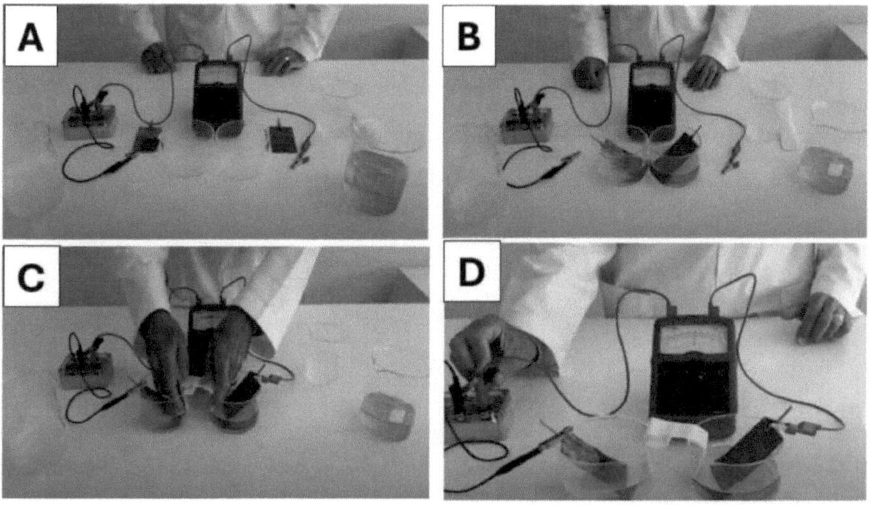


Figure 1: Schematic diagram of experimental set-up.

**The given questionnaire:**

Q1. Which charge carriers are responsible for current flow in the circuit?

__________________________________________________

Q2. What is the direction of the electric current in the ammeter? Can you deduce the direction of movement of the various charge carriers?

__________________________________________________

Q3. What happens at the metal-solution interfaces in the two plates?

__________________________________________________

Q4. What is the role of the ionic bridge

__________________________________________________

## 2. Handling and integration of ICT

After having carried out the same experiment in the previous section, we tried to integrate Information and Communication Technology (ICT) for other five classes: a 2nd year science math class (M2T), 2 physical science classes (P3T and P4T), and 2 life and earth science classes (V3T and V4T), in order to find out the pedagogical effect of integrating ICT, to improve the level of learning among the pupils, and also to motivate them to make better use of technology in education.

In this section we carried out the same experiment shown in Figure 1, and then used the "Daniell" Flash animation digital simulator approved by the Moroccan Ministry of Education.

How do I use the digital simulator?

Several solutions are possible: either download via the link "http://pc1.ma/flash2bac/daniell.swf" or run from a storage medium (USB CD, etc.). The .swf animation file can be played in all web browsers with an up-to-date Adobe Flash Player add-on. It may be incompatible with certain mobile devices. The .exe file contains the animation and a stand-alone flash player.

### *a. Teaching materials:*

Computer, CD-ROM, video projector, Daniell digital stimulator.

The purpose of this animation is to simulate the Pille Daniell experiment, showing the various stages in the operation of this battery.

### *b. Activity stages*

When the digital resource is opened, the first graphical interface is illustrated in figure 2 below, which shows a general diagram of the Daniell bead, in which, after clicking on the magnifying glass (figure 2), the green beads move from the (-) to the (+) pole, making it easier to observe what's happening in the vicinity of the Zn(s)-based (-) electrode, see figure 3.

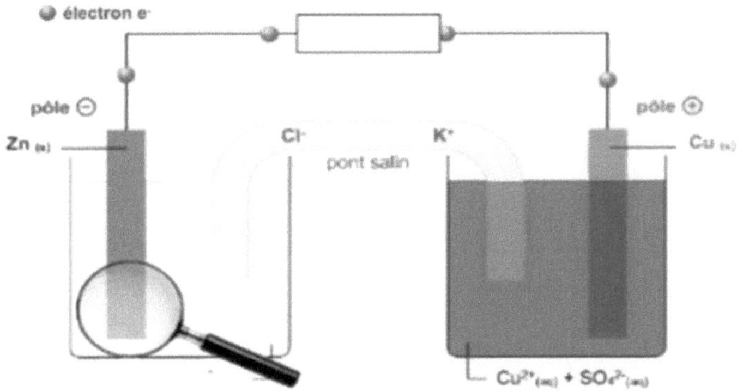


Figure 2: The first graphical interface to appear in digital animation.

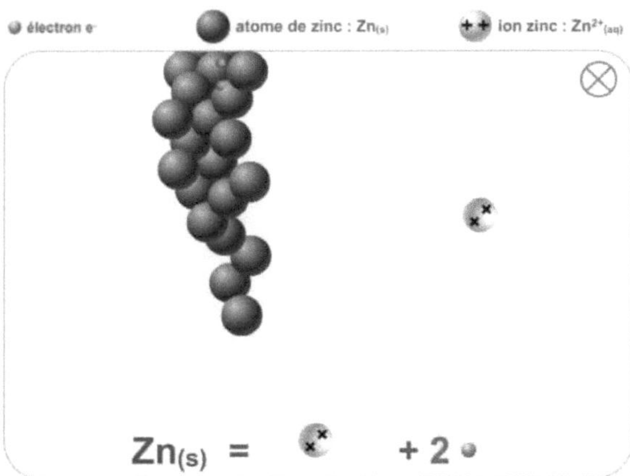


Figure 3: Zn(S) oxidation mechanism.

Clicking on the ® icon (figure3) redisplays figure4, which shows the degradation (oxidation) of the Zn(s) electrode and a deposit on the positive electrode, after which a simple click on the magnifying glass next to the (+) electrode makes it easier to observe what is happening in the vicinity of this electrode, see figure 5.

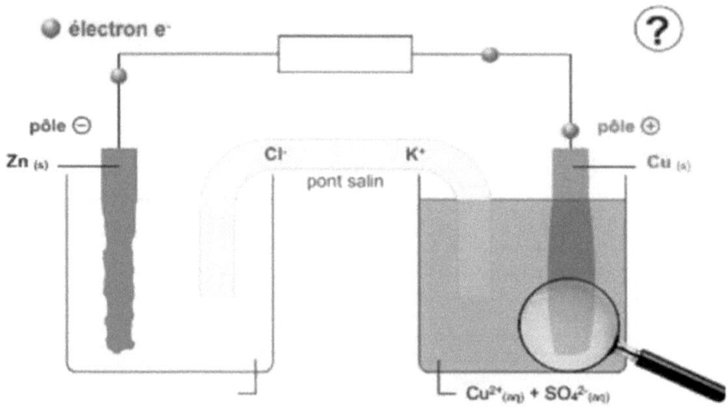


Figure 4: Degradation (oxidation) of the Zn(s) electrode.

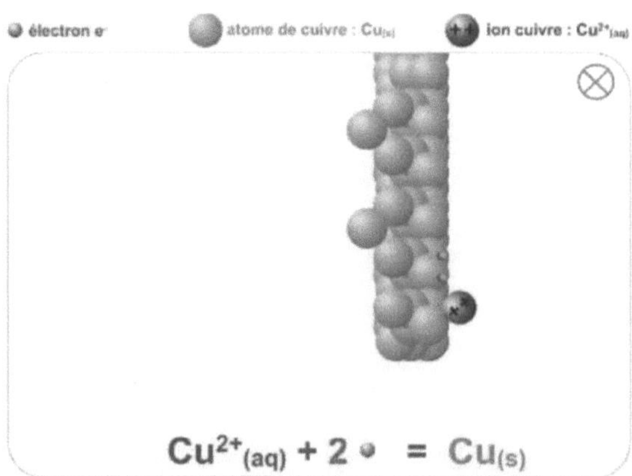


Figure 5: Cu reduction mechanism$^{2+}$

Clicking on the ® icon (figure 5) redisplays the main animation interface, then clicking on the salt bridge section (figure 4) displays what's happening at the salt bridge, showing the mobility of positive and negative ions ($K^+$ (aq) and $Cl^-$ (aq)) in the potassium chloride solution, see figure 6. Figure 7 shows the final state of each electrode at the end of this numerical activity.

Figure 6: Inside the salt bridge

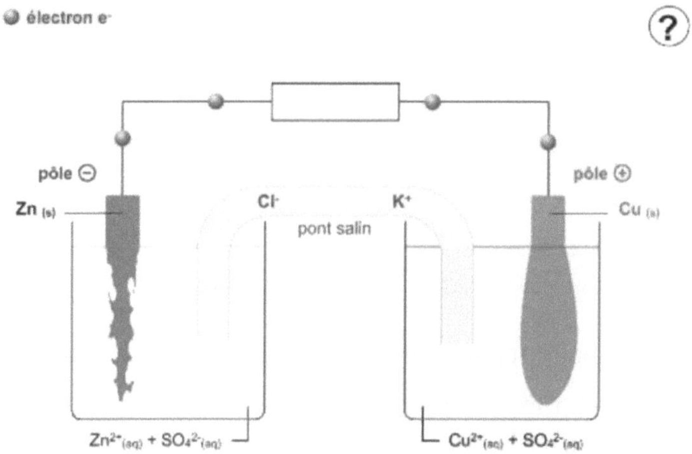


Figure 7: Electrode states at the end of simulation

## II. Analysis of results

The high school chemistry curriculum has three parts: Rapid and slow transformations of a chemical system, Non-total transformations of a chemical system, and Direction of evolution of a chemical system. We have chosen to study the contribution of ICT to the lesson on Spontaneous transformations in batteries and energy production. The questionnaire results obtained in the two studies are classified by class and by the percentage of responses to each question.

### 1) Experimental results

The percentage of answers for each question is calculated by the following expression %= 100 x number of correct answers + the total number of students in each class. The results of each question for all classes are shown in Table 1.

Table 1: Results for each question.

| **The questions** | **The classes** | | | | |
|---|---|---|---|---|---|
| | **M1** | **P1** | **P2** | **V1** | **V2** |
| **Q1** | 67 | 55 | 57 | 41 | 50 |
| **Q2** | 68 | 53 | 61 | 43 | 49 |
| **Q3** | 70 | 63 | 59 | 43 | 48 |
| **Q4** | 71 | 60 | 61 | 46 | 50 |

The results for the science match (M1) options show that students answered all questions with a higher percentage than for the other options, all percentages exceeding 67% and the maximum percentage reaching 71% in the answer to question 4.

Physical science option students (P1 and P2) answered all questions, but with lower percentages than math science students and higher percentages than life and earth science students. Table 1 shows that the minimum percentage was 53%, and the maximum 63%.

Students in the life and earth science option (V1 and V2) answered all the questions, but with lower percentages than the other math and physical science students. The table above shows that the minimum percentage reached the value 41%%, and the maximum percentage was 50%.

Microsoft Excel 2024 was used to represent the results of this study, and Figure 8 shows the graphs obtained.

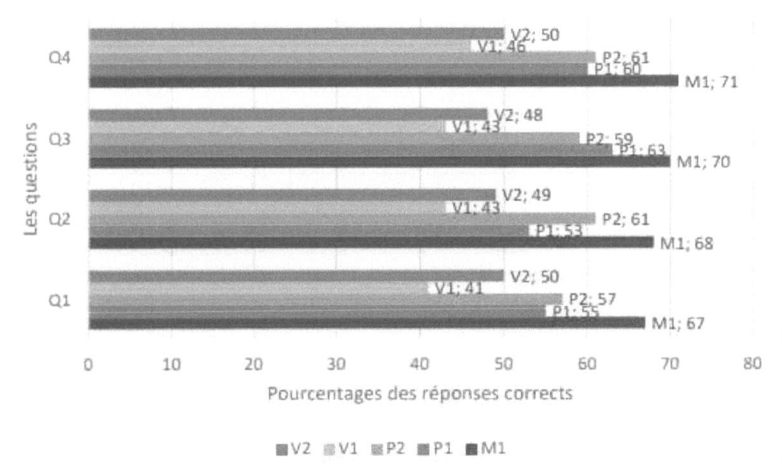


Figure 8: Percentages of correct answers for questions without ICT integration.

The figure above shows that students in the math science class achieve the highest percentages for all questions, followed by the physical science classes, while life and earth students have the lowest percentages for all questions.

### 2) Results achieved through experience and ICT integration

Based on the experiment and the simulation, the same questionnaire was completed, and the results are shown in Table 2.

Table 2: Results for each question with ICT integration.

| The questions | The classes | | | | |
|---|---|---|---|---|---|
| | M2T | P3T | P4T | V3T | V4T |
| **Q1** | 93 | 85 | 86 | 72 | 73 |
| **Q2** | 88 | 82 | 89 | 75 | 72 |
| **Q3** | 91 | 81 | 81 | 71 | 73 |
| **Q4** | 92 | 86 | 83 | 76 | 74 |

The results for match science options (M2T) show that the minimum percentage reached 88% and the maximum 93%, higher than the percentages for other specialties. Students in the physics science option classes (P3T and P4T) answered all questions with higher percentages than students in the life and earth science option classes, ranging from 81% to 89%. Students in the life and earth sciences specialization (V3T and V4T) also achieved significant and acceptable values, ranging from 71% to 76%, although these values remain lower than those of other specializations. We note that all student results have incorporated Information and Communication Technologies (ICT), and experiences are excellent, showing that ICT has played an important and positive role in learning and understanding scientific phenomena.

Microsoft Excel 2024 was used to transform the results of this simulation into a graph. Figure 9 shows the grouped histogram graphs.

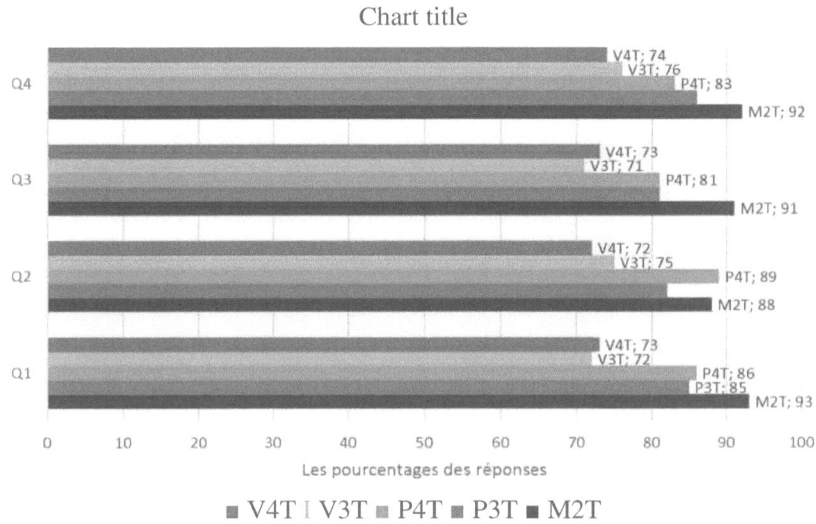


Figure 9: Percentages of correct answers for questions with ICT integration.

The figure above shows that; students in the math science class achieve higher percentages of correct answers for all questions, followed by the physical science classes, while life and earth students have the lowest percentages for all questions. Generally speaking, the values obtained with the integration of ICT are the highest for all specialties.

**3) Comparative study**

In this section, we present a comparative study between the results of the different classes: 5 classes observed only the experiment (M1, P1, P2, V1 and V2), and other 5 classes benefited from the experiment and the integration of numerical simulation (M2T, P3T, P4T, V3T and V4T), the comparative study is illustrated in figure 10.

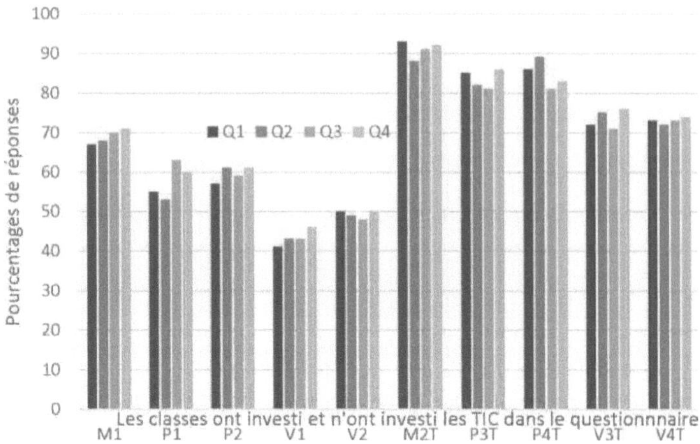


Figure 10: Percentages of correct answers for questions without and with ICT integration.

We note that students who have only benefited from the experience have lower response percentages for all questions and for all subjects, with these percentages decreasing more for life and earth sciences classes.

The comparative study shows that the response percentages increase for all streams that have benefited from experimenting with and integrating Information and Communication Technology (ICT), indicating that ICT plays a very important role in learners' learning and showing things we can't show in experiments like the direction of electron flow, current flow, and the direction of movement of cations and anions in solutions^, these clues can solve many concepts and problems for learners.

## Conclusion

We've tried to highlight the contribution of the new Information and Communication Technologies in Education (ICT), and we've also shown and explained how these techniques could contribute to the necessary changes in learners' learning alongside practical chemistry work, while highlighting the obstacles hindering their learning. ICT techniques propose some solutions to resolve and avoid the learning crisis among learners from which most science classes suffer, precisely in Moroccan high schools. What's more, we've noticed that thanks to ICT, students of all options have become more eager and motivated to learn, it increases response rates for all specialties, these high achievers have also become more acquire new knowledge, they've become interact more and they aspire to become a productive and interactive power in other lessons, reducing academic failure. Generally, the use of ICT techniques in high school teaching has several advantages such as, presenting lessons efficiently and simply, promoting interactive learning, optimizing performance and improving motivation, interaction between teacher and learner in all situations.

## References

[1] D. S. Niederhauser and D. L. Lindstrom, "Instructional Technology Integration Models and Frameworks: Diffusion, Competencies, Attitudes, and Dispositions," in *Second Handbook of Information Technology in Primary and Secondary Education*, J. Voogt, G. Knezek, R. Christensen, and K.-W. Lai, Eds. in Springer International Handbooks of Education. Cham: Springer International Publishing, 2018, pp. 335-355. doi: 10.1007/978-3-319-71054- 9_23.

[2] I. Backfisch, A. Lachner, K. Stürmer, and K. Scheiter, "Variability of teachers' technology integration in the classroom: A matter of utility!," *Computers & Education*, vol. 166, pp. 104159, Jun. 2021, doi: 10.1016/j.compedu.2021.104159.

[3] C. Norris, T. Sullivan, J. Poirot, and E. Soloway, "No Access, No Use, No Impact: Snapshot Surveys of Educational Technology In K#x2013;12," *Journal of Research on Technology in Education*, vol. 36, no. 1, pp. 15-27, Sep. 2003, doi: 10.1080/15391523.2003.10782400.

[4] European Commission. Directorate General for Communications Networks, Content and Technology, Deloitte, and Ipsos MORI, *2nd survey of schools: ICT in education: objective 1: benchmark progress in ICT in schools, final report.* LU: Publications Office, 2019. Accessed: Aug. 21, 2024. [Online]. Available: https://data.europa.eu/doi/10.2759/23401

[5] B. R. Belland, "Using the theory of habitus to move beyond the study of barriers to technology integration," 2009.

[6] V. W. Vongkulluksn, Kui Xie, and M. A. Bowman, "The role of value on teachers' internalization of external barriers and externalization of personal beliefs for classroom technology integration," *Computers & Education*, vol. 118, pp. 70-81, 2018, doi: https://doi.Org/10.1016/j.compedu.2017.11.009.

[7] R. Christensen and G. Knezek, "Instruments for Assessing the Impact of Technology in Education," *Computers in the Schools*, vol. 18, no. 2-3, pp. 525, 2001, doi: https://doi.org/10.1300/J025v18n02_02.

[8] G. Knezek and Rhonda ,Christensen, "Extending the will, skill, tool model of technology integration: adding pedagogy as a new model construct," *J Comput High Educ*, vol. 28, pp. 307-325, 2016.

[9] D. L. Lowther, F. A. Inan, J. D. Strahl, and M. Ross, "Does technology integration 'work' when key barriers are removed?," *Educational Media International*, vol. 45, no. 3, pp. 195-213, 2008, doi: . https://doi.org/10.1080/09523980802284317.

[10] B. Su, "Effective Technology Integration: Old Topic, New Thoughts," *International Journal of Education and Development using ICT*, Mar. 2009, Accessed: Aug. 21, 2024. [Online]. Available: https://www.semanticscholar.org/paper/Effective-Technology-Integration%3A-Old-Topic%2C-New-Su/86427ffec458a91fff73db941a72b0adc7a792a0

[11] M. Bowman, V. Vongkulluksn, Z. Jiang, and K. Xie, "Teachers' exposure to professional development and the quality of their instructional technology use: The mediating role of teachers' value and ability beliefs," *Journal of Research on Technology in Education*, vol. 54, pp. 188-204, May 2022, doi: 10.1080/15391523.2020.1830895.

[12] P. A. Ertmer, Ottenbreit-Leftwich,Anne T, Anne T. Ottenbreit-Leftwich b, Olgun Sadik b, Emine Sendurur c, Polat Sendurur, Olgun Sadik, Olgun, Sendurur, Emine, and Polat Sendurur c, "Teacher beliefs and technology integration practices: A critical relationship," *Computers & Education*, vol. 59, pp. 423-435, 2012, doi: https://doi.org/10.1016Zj.compedu.2012.02.001.

[13] D. Ruggiero and C. J. Mong, "The Teacher Technology Integration Experience: Practice and Reflection in the Classroom," *Journal of Information Technology Education: Research*, vol. 14, pp. 161-178, May 2015, Accessed: Aug. 13, 2024. [Online]. Available: https://www.informingscience.org/Publications/2227

[14] N. Barrett, J. S. Butler, and E. F. Toma, "Do Less Effective Teachers Choose Professional Development Does It Matter?," *Eval Rev*, vol. 36, no. 5, pp. 346374, Oct. 2012, doi: 10.1177/0193841X12473304.

[15] K. Xie, M. K. Kim, S.-L. Cheng, and N. C. Luthy, "Teacher professional development through digital content evaluation," *Education Tech Research Dev*, vol. 65, no. 4, pp. 1067-1103, Aug. 2017, doi: 10.1007/s11423-017-9519-0.

[16] S.-L. Cheng, L. Lu, K. Xie, and V. W. Vongkulluksn, "Understanding teacher technology integration from expectancy-value perspectives," *Teaching and TeacherEducation*, vol. 91, p. 103062, May 2020, doi: 10.1016/j.tate.2020.103062.

[17] V. W. Vongkulluksn, K. Xie, and M. A. Bowman, "The role of value on teachers' internalization of external barriers and externalization of personal beliefs for classroom technology integration," *Computers & Education*, vol. 118, pp. 70-81, Mar. 2018, doi: 10.1016/j.compedu.2017.11.009.

[18] P. A. Ertmer and A. T. Ottenbreit-Leftwich, "Teacher Technology Change: How Knowledge, Confidence, Beliefs, and Culture Intersect," *Journal of Research on Technology in Education*, vol. 42, no. 3, pp. 255-284, Mar. 2010, doi: 10.1080/15391523.2010.10782551.

[19] F. A. Inan and D. L. Lowther, "Factors affecting technology integration in K-12 classrooms: a path model," *Education Tech Research Dev*, vol. 58, no. 2, pp. 137-154, Apr. 2010, doi: 10.1007/s11423-009-9132-y.

[20] J. L. Harris, M. T. Al-Bataineh, and A. Al-Bataineh, "One to One Technology and its Effect on Student Academic Achievement and Motivation," *CONTEMP EDUC TECHNOL*, vol. 7, no. 4, Oct. 2016, doi: 10.30935/cedtech/6182.

[21] K. A. Lawless and J. W. Pellegrino, "Professional Development in Integrating Technology Into Teaching and Learning: Knowns, Unknowns, and Ways to Pursue Better Questions and Answers," *Review of Educational Research*, vol. 77, no. 4, pp. 575-614, Dec. 2007, doi: 10.3102/0034654307309921.

[22] R. S. Davies, "Understanding Technology Literacy: A Framework for Evaluating Educational Technology Integration," *TECHTRENDS TECH TRENDS*, vol. 55, no. 5, pp. 45-52, Sep. 2011, doi: 10.1007/s11528-011-0527- 3.

[23] H. Heflin, J. Shewmaker, and J. Nguyen, "Impact of mobile technology on student attitudes, engagement, and learning," *Computers & Education*, vol. 107, pp. 91-99, Apr. 2017, doi: 10.1016/j.compedu.2017.01.006.

[24] D. Gasevic, J. Jovanovic, A. Pardo, and S. Dawson, "Detecting Learning Strategies with Analytics: Links with Self-reported Measures and Academic Performance," *Learning Analytics*, vol. 4, no. 2, Jul. 2017, doi: 10.18608/jla.2017.42.10.

[25] M. M. Msafiri, D. Kangwa, and L. Cai, "A systematic literature review of ICT integration in secondary education: what works, what does not, and what next?," *Discov Educ*, vol. 2, no. 1, p. 44, Nov. 2023, doi: 10.1007/s44217-023-00070- x.

[26] E. Karahan, S. C. Bilici, and A. Ünal, "Integration of Media Design Processes in Science, Technology, Engineering, and Mathematics (STEM) Education," *EJER,* vol. 15, no. 60, pp. 221-240, Sep. 2015, doi: 10.14689/ejer.2015.60.15.

[27] K. E. Lees *et al*, "NUCare: Advancing research on technological integration for self-management in the aging population," *Nursing Outlook*, vol. 66, no. 2, pp. 121-129, Mar. 2018, doi: 10.1016/j.outlook.2017.10.008.

[28] N. Smeda, E. Dakich, and N. Sharda, "The effectiveness of digital storytelling in the classrooms: a comprehensive study," *Smart Learn. Environ.* vol. 1, no. 1, p. 6, Dec. 2014, doi: 10.1186/s40561-014-0006-3.

[29] S. D. Milner and A. M. Wildberger, "Determining appropriate uses of computers in education," *Computers & Education*, vol. 1, no. 2, pp. 117-123, Jan. 1977, doi: 10.1016/0360-1315(77)90006-9.

[30] R. Raja and P. C. Nagasubramani, "Impact of modern technology in education," *jaar*, pp. S33-S35, May 2018, doi: 10.21839/jaar.2018.v3iS1.165.

[31] A. Haleem, M. Javaid, M. A. Qadri, and R. Suman, "Understanding the role of digital technologies in education: A review," *Sustainable Operations and Computers*, vol. 3, pp. 275-285, 2022, doi: 10.1016/j.susoc.2022.05.004.

[32] V. Varea, G. Gonzalez-Calvo, and A. Garda-Monge, "Exploring the changes of physical education in the age of Covid-19," *Physical Education and Sport Pedagogy*, vol. 27, no. 1, pp. 32-42, Jan. 2022, doi: 10.1080/17408989.2020.1861233.

[33] R. N. Carvalho, C. E. F. Monteiro, and M. N. P. Martins, "Challenges for university teacher education in Brazil posed by the Alpha Generation," *REALIA*, no. 28, p. 61, Jan. 2022, doi: 10.7203/realia.28.21408.

[34] J. Hsu, "Innovative Technologies for Education and Learning: Education and Knowledge-Oriented Applications of Blogs, Wikis, Podcasts, and More," *Int. J. Inf. Commun. Technol. Educ.* pp. 70-89, 2007, Accessed: Aug. 27, 2024. [Online]. Available: https://www.semanticscholar.org/paper/Innovative-Technologies-for-Education-and-Learning%3A-Hsu/3f583d37e74e42f5f66b82c79c8557e1514abc93

[35] R. Grainger, Q. Liu, and S. Geertshuis, "Learning technologies: A medium for the transformation of medical education?," *MedEduc,* vol. 55, no. 1, pp. 23-29, Jan. 2021, doi: 10.1111/medu.14261.

[36] E. Lacka and T. C. Wong, "Examining the impact of digital technologies on students' higher education outcomes: the case of the virtual learning environment and social media," *Studies in Higher Education*, vol. 46, no. 8, pp. 1621-1634, Aug. 2021, doi: 10.1080/03075079.2019.1698533.

[37] T. Daubenfeld and D. Zenker, "A Game-Based Approach to an Entire Physical Chemistry Course," *J. Chem. Educ.* vol. 92, no. 2, pp. 269-277, Feb. 2015, doi: 10.1021/ed5001697.

[38] J. Seibert, C. W. M. Kay, and J. Huwer, "EXPlainistry: Creating Documentation, Explanations, and Animated Visualizations of Chemistry Experiments Supported by Information and Communication Technology To Help School Students Understand Molecular-Level Interactions," *J. Chem. Educ*, vol. 96, no. 11, pp. 2503-2509, Nov. 2019, doi: 10.1021/acs.jchemed.8b00819.

[39] K. Suh, T. Smith, and M. Linhoff, "Leveraging Socially Networked Mobile ICT Platforms for the Last-Mile Delivery Problem," *Environ. Sci. Technol.* vol. 46, no. 17, pp. 9481-9490, Sep. 2012, doi: 10.1021/es301302k.

[40] M. Gawlik-Kobylinska, W. Walkowiak, and P. Maciejewski, "Improvement of a Sustainable World through the Application of Innovative Didactic Tools in Green Chemistry Teaching: A Review," *J. Chem. Educ.* vol. 97, no. 4, pp. 916924, Apr. 2020, doi: 10.1021/acs.jchemed.9b01038.

[41] T. Gupta and R. E. Belford, "Conclusion: Technology Integration in Chemistry Education and Research: What Did We Learn and What Can We Expect Going Forward?", in *ACS Symposium Series*, vol. 1318, T. Gupta and R. E. Belford, Eds, Washington, DC: American Chemical Society, 2019, pp. 281-301. doi: 10.1021/bk-2019-1318.ch018.

[42] R. A. Tigaa and S. L. Sonawane, "An International Perspective: Teaching Chemistry and Engaging Students During the COVID-19 Pandemic," *J. Chem. Educ.* vol. 97, no. 9, pp. 3318-3321, Sep. 2020, doi: 10.1021/acs.jchemed.0c00554.

[43] R. EL-Mernissi *et al*, "Designing of Novel Quinolines Derivatives as Hepatocellular Carcinoma Inhibitors by Using In silico Approaches," *Biointerface Res Appl Chem*, vol. 13, no. 3, pp. 1-15, Jun. 2022, doi: 10.33263/BRIAC133.217.

[44] R. El-Mernissi *et al*, "In silico studies of novel styrylquinoline derivatives as tubulin polymerization inhibitors against human ovarian carcinoma," *Malaysian Journal of Biochemistry & Molecular Biology*, pp. 117-129, 2022.

[45] R. El-Mernissi *et al*, "3D-QSAR, molecular docking, simulation dynamic and ADMET studies on new quinolines derivatives against colorectal carcinoma activity," *Journal of Biomolecular Structure and Dynamics*, vol. 42, no. 7, pp. 3682-3699, May 2024, doi: 10.1080/07391102.2023.2214233.

[46] R. El-Mernissi *et al*, "New compounds based on 1H-pyrrolo[2,3-b] pyridine as potent TNIK inhibitors against colorectal cancer cells. Molecular modeling studies," *Moroccan Journal of Chemistry*, vol. 11, no. 1, pp. 20-33, 2023.

[47] R. EL-Mernissi *et al*, "3D-QSAR, molecular docking, ADMET, simulation dynamic, and retrosynthesis studies on new styrylquinolines derivatives against breast cancer," *Open Chemistry*, vol. 22, no. 1, p. 20240041, Jun. 2024, doi: 10.1515/chem-2024-0041.

[48] R. El-Mernissi *et al*, "3D- QSAR, ADMET, and Molecular Docking Studies for Designing New 1,3,5-Triazine Derivatives as Anticancer Agents," *Egyptian Journal of Chemistry,* vol. 65, no. 132, pp. 9-18, Dec. 2022, doi: 10.21608/ejchem.2022.76000.3715.

[49] L. El Mchichi *et al*, "In silico design of novel Pyrazole derivatives containing thiourea skeleton as anti-cancer agents using: 3D QSAR, Drug-Likeness studies, ADMET prediction and molecular docking," *Materials Today: Proceedings*, vol. 45, pp. 7661-7674, 2021, doi: 10.1016/j.matpr.2021.03.152.

[50] R. El-Mernissi *et al*, "2-Oxoquinoline Arylaminothiazole Derivatives in Identifying Novel Potential Anticancer Agents by Applying 3D-QSAR, Docking, and Molecular Dynamics Simulation Studies," *J. Mex. Chem. Soc.* vol. 66, no. 1, pp. 79-94, Dec. 2021, doi: 10.29356/jmcs.v66i1.1578.

[51] R. EL-Mernissi *et al*, "Design of new 3, 5-disubstituted indole as hematological anticancer agents using 3D-QSAR, molecular docking and drug-likeness studies," *Materials Today: Proceedings*, vol. 45, pp. 7608-7614, 2021, doi: 10.1016/j.matpr.2021.03.080.

[52] A. Nejjari and I. Bakkali, "L'usage des TIC à l'école marocaine: état des lieux et perspectives," *Hermès*, vol. n° 78, no. 2, p. 55, 2017, doi: 10.3917/herm.078.0055.

[53] M. Chekour, M. Laafou, and R. Janati-Idrissi, "Factors influencing concept acquisition in electricity," https://adjectif.net, Aug. 2015, Accessed: Jul. 26, 2024. [Online]. Available: https://adjectif.net//spip.php?article354

FSC
www.fsc.org
MIX
Papier aus verantwortungsvollen Quellen
Paper from responsible sources
FSC® C105338

Printed by Books on Demand GmbH, Norderstedt / Germany